MRI Atlas of Pituitary Pathology

MRI Atlas of Pituitary Pathology

Kevin M. Pantalone, DO
Cleveland Clinic, Cleveland, OH, USA

Stephen E. Jones, MD, PhD
Cleveland Clinic, Cleveland, OH, USA

Robert J. Weil, MD
Geisinger Health System, PA, USA

Amir H. Hamrahian, MD
Chief, Endocrinology, Professor of Medicine,
Cleveland Clinic Abu Dhabi, Abu Dhabi, UAE;
Staff Endocrinologist, Cleveland Clinic,
Cleveland, OH, USA

AMSTERDAM • BOSTON • HEIDELBERG • LONDON
NEW YORK • OXFORD • PARIS • SAN DIEGO
SAN FRANCISCO • SINGAPORE • SYDNEY • TOKYO

Academic Press is an imprint of Elsevier

Academic Press is an imprint of Elsevier
32 Jamestown Road, London NW1 7BY, UK
525 B Street, Suite 1800, San Diego, CA 92101-4495, USA
225 Wyman Street, Waltham, MA 02451, USA
The Boulevard, Langford Lane, Kidlington, Oxford OX5 1GB, UK

ISBN: 978-0-12-802577-2

British Library Cataloguing-in-Publication Data
A catalogue record for this book is available from the British Library

Library of Congress Cataloging-in-Publication Data
A catalog record for this book is available from the Library of Congress

For information on all Academic Press publications
visit our website at http://store.elsevier.com/

Typeset by MPS Limited, Chennai, India
www.adi-mps.com

Printed and bound in USA

**Working together
to grow libraries in
developing countries**

www.elsevier.com • www.bookaid.org

Contents

Atlas of Pituitary Imaging

Introduction	1
Normal Pituitary Gland Anatomy	2
MRI Interpretation	2
Normal Pituitary Gland and Sella Anatomy on MRI	4
Pituitary Adenoma	7
Pituitary Microadenoma	7
Pituitary Macroadenoma	8
Pituitary Macroadenoma	9
Pituitary Macroadenoma with Stalk Deviation	9
Pituitary Macroadenoma with Mild Superior Displacement on the Optic Chiasm	10
Large Invasive Pituitary Macroadenoma	11
Patient with a Giant Invasive Prolactinoma	12
Pituitary Adenoma with Cavernous Sinus Invasion	13
Cystic Pituitary Macroadenoma	14
Atypical/Ectopic Adenomas	15
Atypical Adenoma	15
Ectopic Adenoma	15
Cystic Lesions	17
Rathke Cleft Cyst	17
Arachnoid Cyst	18
Dermoid/Epidermoid Cysts	19
Dermoid Cyst	19
Epidermoid Cyst	20
Pituicytoma	21
Chordoma	22
Empty Sella	23
Partial Empty Sella	25
Germinoma	25
Craniopharyngioma	26
Meningioma	27
Pituitary Hyperplasia	29

Anatomic Variations   30
Pituitary Stalk Transection   32
Vascular Lesions (Aneurysm)   33
Infiltrative Disorders   34
Lymphocytic Hypophysitis   34
Langerhans Cell Histiocytosis   35
Neurosarcoidosis   36
Wegener's Granulomatosis   37
Hemochromatosis   38
Metastases   39
Primary Pituitary Lymphoma   40
Pituitary Infection/Abscess   40
Postoperative Pituitary Imaging   41
Visible Fat Pad   41
Sellar Remodeling Post-Pituitary Surgery   41
Apoplexy   42
Hemorrhage   42
Infarction (Non-Hemorrhagic)   45
Volume Averaging   46
CT Sella   47
References   49

Index   51

Atlas of Pituitary Imaging

INTRODUCTION

The basic understanding and interpretation of magnetic resonance imaging (MRI) is important for many clinicians outside of the field of radiology, especially for endocrinologists who may have received limited formal training in such areas. Combining a detailed clinical history, lab data, and images may provide the endocrinologist with an upper hand in evaluating patients with pituitary disorders. Accordingly, it is important that endocrinologists be familiar with reviewing pituitary images. The purpose of this atlas is to provide readers with a simple approach for the evaluation of pituitary images. This review does not preclude timely review of pituitary cases with a neuroradiologist. It needs to be emphasized that the best patient outcome usually requires a multidisciplinary team of endocrinologists, radiation oncologists, and neurosurgeons.

MRI is the imaging modality of choice to evaluate pituitary disorders since it provides a detailed anatomy of the pituitary gland and surrounding structures, particularly the soft tissues. MRI utilizes the application of a strong static magnetic field to align some of the protons within water molecules, and radiofrequency fields to alter their alignments and produce a signal that is detectable outside the body. Various characteristics of the subsequent relaxation process of these protons are used to construct detailed images of the anatomy. Since different tissues have different relaxation properties, images can be generated with exquisite tissue contrast. Furthermore, different MRI sequences can produce different styles of tissue contrast – for example, T1-weighted or T2-weighted images.

Computed tomography (CT) of the sella will be discussed briefly since an MRI may not be appropriate in all patients (e.g., those with a cardiac pacemaker). At the same time, CT may provide the necessary imaging for some patients at lower cost, provide high-resolution and sensitive visualization of any calcifications associated with lesions, and provide superior distinction of osseous anatomy to help guide medical decision-making or surgical procedures.

MRI Atlas of Pituitary Pathology. DOI: http://dx.doi.org/10.1016/B978-0-12-802577-2.00001-3

NORMAL PITUITARY GLAND ANATOMY

The pituitary gland projects from the inferior aspect of the hypothalamus, maintaining a functional connection via the infundibulum (pituitary stalk) (Figure 1.1). It resides in a saddle-like bone cavity referred to as the sella turcica, and is covered by a dural fold (diaphragm sella). Superior to the pituitary gland is the suprasellar cistern containing the optic chiasm. Lateral to the sella turcica are the cavernous sinuses (a large, thin-walled venous plexus), which contain the internal carotid artery and cranial nerves III (oculomotor), IV (trochlear), V [trigeminal, branches V1 (ophthalmic) and V2 (maxillary)], and VI (abducens). See Figure 1.1 for a schematic representation.

The pituitary gland is comprised of anterior and posterior lobes. The intermediate lobe, located between the anterior and posterior pituitary, is rudimentary in humans, and usually absent after birth. The anterior pituitary secretes hormones under the influence of the hypothalamus, the main hormones being growth hormone (GH), thyroid-stimulating hormone (TSH), adrenocorticotropic hormone (ACTH), prolactin (PRL), and the gonadotropins [luteinizing hormone (LH) and follicle-stimulating hormone (FSH)]. The posterior pituitary secretes oxytocin and antidiuretic hormone.

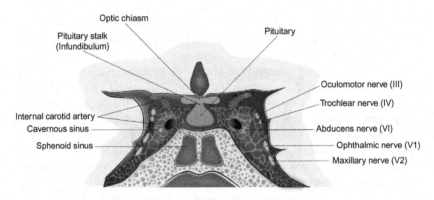

FIGURE 1.1 Schematic of normal pituitary gland anatomy.

MRI INTERPRETATION

I. The first step in interpreting MR imaging is to correctly identify T1- and T2-weighted (Figure 1.2A and B) and pre- and postcontrast images (Figure 1.3A−D). These different images complement one another and are both useful in characterizing sellar anatomy and pathology.

T1-weighted images can easily differentiate fat and water from other common intracranial tissues: fat appears brighter, water appears darker. T2-weighted images have a similar capacity but with opposite features: fat appears darker, and water appears brighter. Thus, within the head, the most common T2-weighted bright object is the cerebrospinal fluid (CSF), and this is particularly evident in the cisterns surrounding the sella. In practice, directing one's attention immediately to the lateral cerebral ventricles will easily assist the viewer in differentiating T1- and T2-weighted images; the CSF will appear black on T1-weighted images (Figure 1.2A) and white on T2-weighted (Figure 1.2B) images. Both T1- and T2-weighted images contrast the gray and white matter. Figure 1.2A was obtained from a previous publication by our group [1].

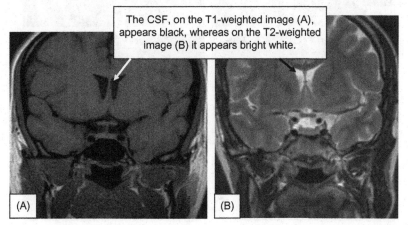

The CSF, on the T1-weighted image (A), appears black, whereas on the T2-weighted image (B) it appears bright white.

(A)   (B)

FIGURE 1.2   (A) Coronal T1-weighted image without contrast, and (B) coronal T2-weighted image.

II. After distinguishing T1- and T2-weighted images, the next step is to identify the pre- and postcontrast T1-weighted images. For physicians with less familiarity with pituitary imaging, looking at the nasal conchae is an easy way to differentiate between pre- and postcontrast images. In precontrast images (Figure 1.3A and C), the nasal conchae have intensity similar to the brain gray matter (isointense); however, on postcontrast images (Figure 1.3B and D), the nasal conchae appear brighter than the gray matter (hyperintense). Note that the pachymeninges (dura mater) and falx also show conspicuous brightening after gadolinium administration. MRI contrast agents (which use a water-soluble gadolinium chelate) are used to increase the ability to discriminate between body tissues and fluids by altering their relaxation times (because of their underlying degree of vascularity). In addition, enhancement in the brain often reflects breakdown of the blood—brain barrier. Enhancement is most vivid on a T1-weighted sequence; although T2-weighted sequences enhance to a minor degree, postcontrast sequences are not obtained with this weighting.

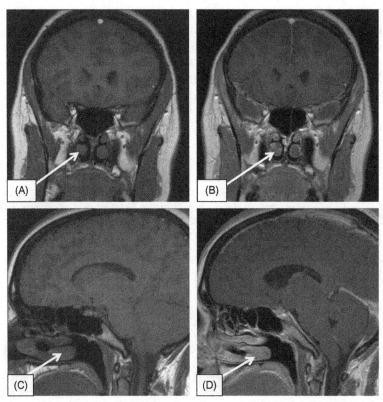

FIGURE 1.3   (A) T1-weighted coronal image precontrast, (B) T1-weighted coronal image postcontrast, (C) T1-weighted sagittal image precontrast, and (D) T1-weighted sagittal image postcontrast.

Normal Pituitary Gland and Sella Anatomy on MRI (Figure 1.4)

The pituitary stalk typically is less than 4 mm in width (Figure 1.5B, thick white arrow); it is usually narrower near the optic chiasm (~2 mm) and widest (~4 mm) at its pituitary insertion [2]. The pituitary stalk may be centrally located or it may be slightly deviated in patients with no known pituitary pathology. A slightly deviated or tilted pituitary stalk has been reported in up to 50% of subjects undergoing pituitary MRI [3]. A stalk thickness greater than 4 mm is usually considered pathologic, which needs to be evaluated in the context of the clinical picture.

The size of the pituitary gland (Figure 1.5B, thick white dashed arrow) varies with both age and sex. On average, the pituitary is 3–8 mm in size (height), and is generally larger in females than in males. The height increases during adolescence due to the normal physiological hypertrophy that occurs during puberty. In both sexes, the maximum pituitary gland height

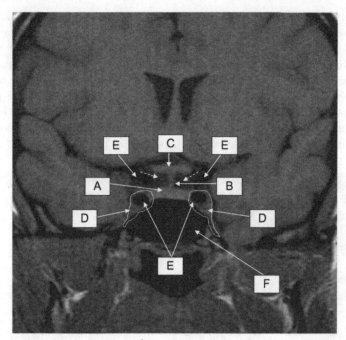

FIGURE 1.4 T1-weighted coronal image without contrast: (A) pituitary gland, (B) pituitary stalk, (C) optic chiasm, (D) cavernous sinus, (E) internal carotid arteries, and (F) sphenoid sinus. The suprasellar cistern is commonly mentioned in radiology reports; it is the cerebrospinal fluid-filled space located above the sella turcica and under the hypothalamus. It contains the optic chiasm, the infundiblar stalk, and the circle of Willis (see white dashed arrows).

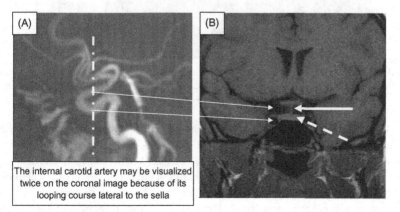

The internal carotid artery may be visualized twice on the coronal image because of its looping course lateral to the sella

FIGURE 1.5 (A) Time-of-flight MRA, sagittal view of internal carotid arteries, and (B) coronal T1-weighted image without contrast.

is usually observed in those within the 20- to 29-year age group. In almost all men, the pituitary height is less than 7 mm, and in women it is less than 8 mm [4]. However, the height of the gland can be up to 9 mm in some young female patients, having a convex superior surface [5]. There is also a slight increase in size during the sixth decade in females, secondary to the rise in gonadotropins. The most striking physiological changes are seen during pregnancy, when the gland progressively enlarges, usually reaching a maximal height of 10 mm immediately after giving birth. On occasion, a height upwards of 12 mm may be observed in the immediate postpartum period [6].

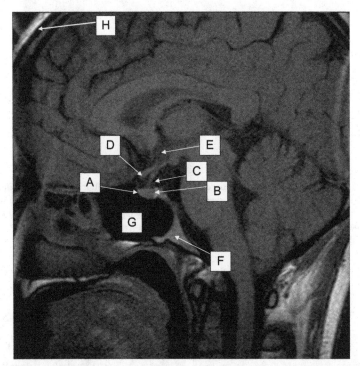

FIGURE 1.6 T1-weighted sagittal image without contrast: (A) anterior pituitary; (B) posterior pituitary; (C) pituitary stalk; (D) optic chiasm; (E) third ventricle, showing adjacent hypothalamus; (F) clivus; (G) sphenoid sinus; and (H) outer table of the calvarium.

Cortical bone and air do not sufficiently mobilize protons to give an MRI signal, and thus they appear black on both T1- and T2-weighted images. However, the central aspects of bones may contain a significant amount of marrow (consisting mainly of fat cells), which appears hyperintense on T1-weighted images. A good example is seen within the clivus (F), due to its commonly high fat content within the bone marrow. This feature of bone marrow is in contrast to the cortical margins of the skull (H), which appear black. Note that the sphenoid sinus (G), which is filled with air, also appears black (Figure 1.6).

PITUITARY ADENOMA

A pituitary adenoma less than 1 cm is referred to as a microadenoma; otherwise, it is referred to as a macroadenoma.

PITUITARY MICROADENOMA

Pituitary adenomas may appear isointense or hypointense on noncontrast T1-weighted imaging when compared to normal anterior pituitary tissue. Since most pituitary tumors are less vascular than normal pituitary tissue, they appear hypointense compared to the surrounding pituitary tissue during postcontrast studies (Figure 1.7A–D). The microadenoma on the T2-weighted image (Figure 1.7E) is not isointense compared to CSF (which appears bright), suggesting it is not cystic in nature (small white arrow).

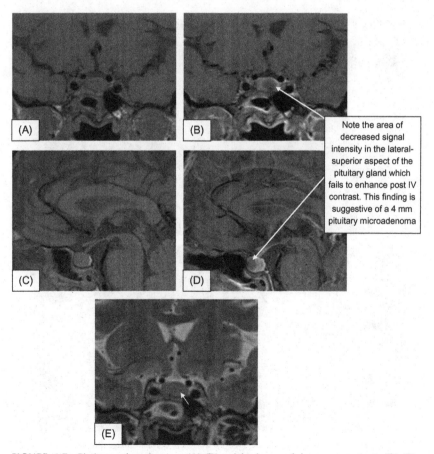

Note the area of decreased signal intensity in the lateral-superior aspect of the pituitary gland which fails to enhance post IV contrast. This finding is suggestive of a 4 mm pituitary microadenoma

FIGURE 1.7 Pituitary microadenoma: (A) T1-weighted coronal image precontrast, (B) T1-weighted coronal image postcontrast, (C) T1-weighted sagittal image precontrast, (D) T1-weighted sagittal image postcontrast, and (E) T2-weighted coronal image.

PITUITARY MACROADENOMA

A nonenhancing mass lesion is noted to be present centrally and to the left of the sella, crossing the midline and measuring 1.6 cm × 1.4 cm × 1.3 cm. The infundibulum is not well visualized in all images, but can be appreciated slightly deviated toward the right on the coronal image (Figure 1.8A, white arrow). The sellar mass abuts the optic chiasm but causes no mass effect. The normal pituitary gland is compressed and located to the far lateral right of the sella (Figure 1.8B, white dashed arrow). Part of the mass has high signal intensity during T2-weighted imaging (Figure 1.8E, white arrow). The patient underwent surgery and was found to have a pituitary adenoma with cystic/necrotic degeneration.

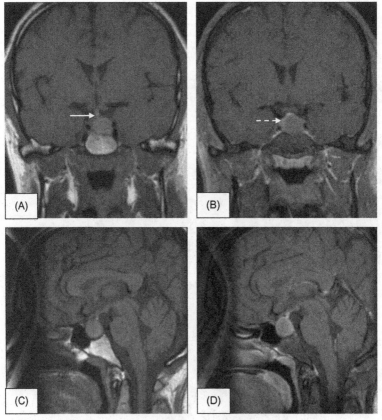

FIGURE 1.8 Pituitary macroadenoma: (A) T1-weighted coronal image precontrast, (B) T1-weighted coronal image postcontrast, (C) T1-weighted sagittal image precontrast, (D) T1-weighted sagittal image postcontrast.

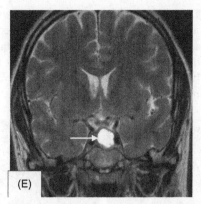

FIGURE 1.8   Pituitary macroadenoma: (E) T2-weighted coronal image. Inherent T2 signal intensity within the mass suggested a cystic lesion.

PITUITARY MACROADENOMA

Figure 1.9 shows a T1-weighted postcontrast coronal image of a pituitary macroadenoma measuring 1.1 cm × 1.1 cm × 1.3 cm, with the normal gland compressed superiorly and to the right. The pituitary stalk cannot be well visualized in this image. Note the sellar floor slopes downward to the right.

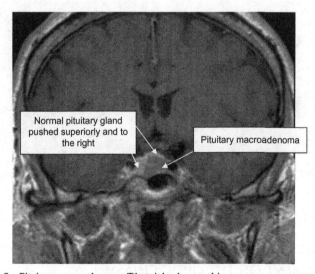

FIGURE 1.9   Pituitary macroadenoma: T1-weighted coronal image postcontrast.

PITUITARY MACROADENOMA WITH STALK DEVIATION

A nonenhancing mass is present within the left aspect of the pituitary gland; it measures 1.5 cm × 1.4 cm × 1.4 cm. The stalk is mildly deviated to the right (white arrow). The imaging characteristics are consistent with that of a macroadenoma. The adenoma is not significantly enhanced after the administration of IV contrast, whereas the area of normal pituitary tissue displaced to the right, observed at the site of stalk termination, is noted to be mildly enhanced (Figure 1.10B). The sellar floor slopes downward to the left, and the mass does not abut or compress the optic chiasm.

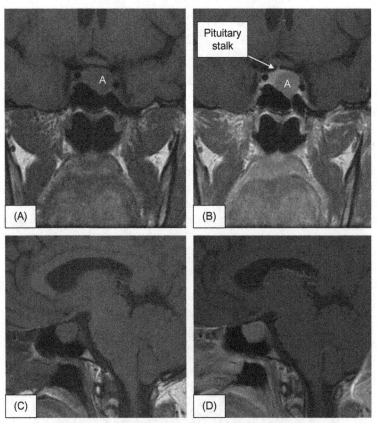

FIGURE 1.10   Pituitary macroadenoma with stalk deviation: (A) T1-weighted coronal image precontrast, (B) T1-weighted coronal image postcontrast, (C) T1-weighted sagittal image precontrast, and (D) T1-weighted sagittal image postcontrast.

PITUITARY MACROADENOMA WITH MILD SUPERIOR DISPLACEMENT ON THE OPTIC CHIASM

Figure 1.11 shows some subtle normal pituitary tissue seen as a slightly increased enhancement along the left lateral portion of the sellar mass (thick white arrow). Mild superior displacement of the optic chiasm is present (thin white arrow). The bright white areas seen bilateral to the inferior aspect of the mass are a normal enhancement of the cavernous sinuses (dashed white arrows).

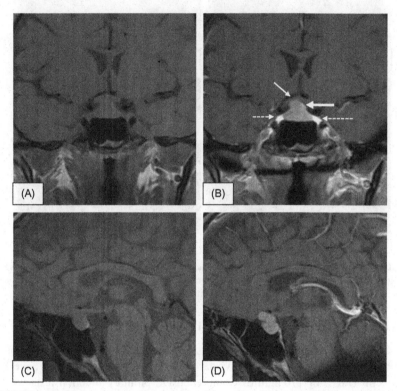

FIGURE 1.11   Pituitary macroadenoma with mild superior displacement of the optic chiasm: (A) T1-weighted coronal image precontrast, (B) T1-weighted coronal image postcontrast, (C) T1-weighted sagittal image precontrast, and (D) T1-weighted sagittal image postcontrast.

LARGE INVASIVE PITUITARY MACROADENOMA

Figure 1.12 shows an image from a 54-year-old man with a large invasive prolactinoma that has grown inferiorly without compromising the optic chiasm. The pituitary stalk is deviated to the left (thin arrow). The tumor heterogeneously enhances postcontrast (thick arrow) and grossly invades (obliterates) the sphenoid sinus, which is best seen on sagittal images.

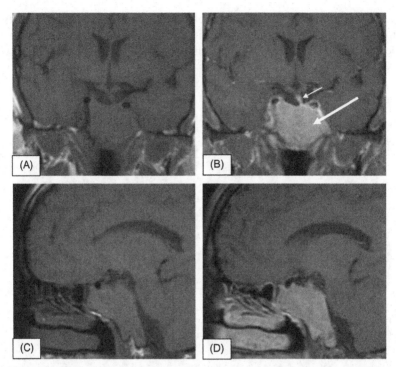

FIGURE 1.12 Large invasive pituitary macroadenoma: (A) T1-weighted coronal image precontrast, (B) T1-weighted coronal image postcontrast, (C) T1-weighted sagittal image precontrast, and (D) T1-weighted sagittal image postcontrast.

PATIENT WITH A GIANT INVASIVE PROLACTINOMA

A 46-year-old man with a giant invasive sellar mass with expansion and destruction, causing a bilateral visual field defect, was found to have a prolactin (PRL) of 110 ng/mL. Comprehensive pituitary function tests were only significant for central hypogonadism and low IGF-1 (Figure 1.13).

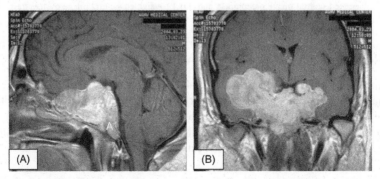

FIGURE 1.13 Patient with a giant invasive prolactinoma: (A) Preoperative T1-weighted sagittal image postcontrast, and (B) preoperative T1-weighted coronal image postcontrast.

Patients with a giant pituitary adenoma and elevated PRL levels, suggestive of a stalk effect (pituitary stalk compression resulting in mild to moderate hyperprolactinemia via interfering with normal dopamine inhibition of PRL secretion), should always undergo serum dilution if a two-step PRL assay is not being utilized. PRL elevations secondary to a stalk effect are usually less than 100 ng/mL and almost always less than 200 ng/mL. The "hook effect" can be observed in the setting of large PRL-secreting tumors (>3 to 4 cm), where large quantities of antigen (PRL) in an immunoassay impair antigen–antibody binding, resulting in erroneously low (normal-range or only mildly elevated antigen determinations). Simply stated, the extremely high levels of PRL overwhelm the reagent antibody of the assay. Figure 1.13 was obtained from a previous publication by our group [7]. Unfortunately, this patient was sent to surgery prior to further evaluation of the serum PRL. After surgery, the patient's serum PRL level was found to be 13,144 ng/mL using our institution's two-step assay that eliminates the hook effect.

PITUITARY ADENOMA WITH CAVERNOUS SINUS INVASION

In Figure 1.14A, note the invasion of the left cavernous sinus by the macroadenoma, which fails to enhance postcontrast, unlike that of the adjacent normal pituitary tissue.

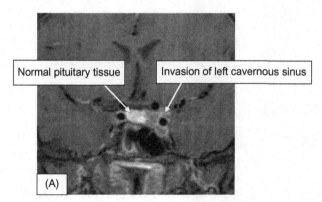

FIGURE 1.14   Pituitary adenoma with cavernous sinus invasion: (A) T1-weighted coronal image postcontrast.

There is no consensus on defining cavernous sinus invasion, but one rule-of-thumb approach is to draw the middle/median intercarotid line (Figure 1.14B). Frank cavernous sinus invasion is said to be present if the adenoma surpasses the middle/median intercarotid line, which suggests at least 50% invasion. Both cuts of the internal carotid artery may not appear on the same image that best demonstrates the invasion; thus, this may often need to be visually approximated after viewing all of the images.

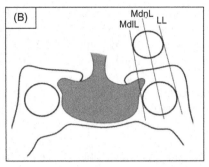

FIGURE 1.14 (B) schematic, cavernous sinus invasion [MdlL (medial line), MdnL (median line), and LL (lateral line)].

CYSTIC PITUITARY MACROADENOMA (FIGURE 1.15)

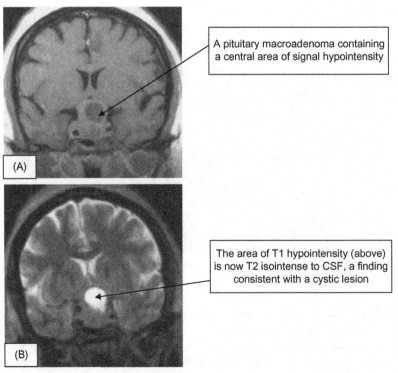

A pituitary macroadenoma containing a central area of signal hypointensity

The area of T1 hypointensity (above) is now T2 isointense to CSF, a finding consistent with a cystic lesion

FIGURE 1.15 Cystic pituitary macroadenoma: (A) T1-weighted coronal image postcontrast, and (B) T2-weighted coronal image.

ATYPICAL/ECTOPIC ADENOMAS

Some pituitary adenomas may have an atypical location, appearance, or presentation. A tissue diagnosis may at times be necessary to differentiate them from other rare pathologies.

Atypical Adenoma

A 41-year-old woman presented with severe headaches localized behind the left eye and on the left side of her face. On exam she had partial third nerve palsy along with mild left eye proptosis. MRI revealed a left-sided sellar mass invading the left cavernous sinus (Figure 1.16). The relative lack of enhancement of the mass after administration of IV contrast (white arrows), in relation to the robust enhancement of that of the normal pituitary gland (black arrow), was suggestive of a pituitary adenoma. However, the presentation with left facial pain, eye pain, partial third nerve palsy, and mild proptosis was suggestive of a nonadenomatous lesion. Pathology revealed the mass to be a silent ACTH adenoma. At surgery, the lesion was located wholly within the cavernous sinus and did not involve the pituitary gland.

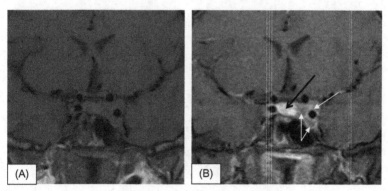

FIGURE 1.16 Atypical adenoma: (A) T1-weighted coronal image precontrast, and (B) T1-weighted coronal image postcontrast.

Ectopic Adenoma

A 39-year-old man was referred for evaluation of hypercortisolemia. The patient was diagnosed with ACTH-dependent cyclic Cushing syndrome. The patient underwent trans-sphenoidal surgery, during which a 5 mm firm, round, midline sphenoid sinus lesion was identified and resected (Figure 1.17).

The lesion was found to be a pituitary adenoma that stained diffusely with ACTH antibodies. The patient developed adrenal insufficiency postop, and has been in remission since then. Explorations of the sella and pituitary did not reveal any abnormalities. Figure 1.17A shows a T1-weighted coronal image without contrast. The sellar anatomy appears normal with an unremarkable pituitary gland. Figure 1.17B shows a T1-weighted coronal image without contrast in a more anterior plane. Note the small soft tissue mass visible in the sphenoid sinus anterior to the sella (thin arrow). A T1-weighted sagittal image postcontrast is shown in Figure 1.17C. Note the normal enhancing pituitary gland (thick arrow) and the soft-tissue mass anteriorly in the sphenoid sinus (thin arrow). The sphenoid sinus lesion (thin arrow) was interpreted by the neuroradiologist as being consistent with a mucosal polyp. The clinical case and pictures were obtained from a previous publication by our group [8].

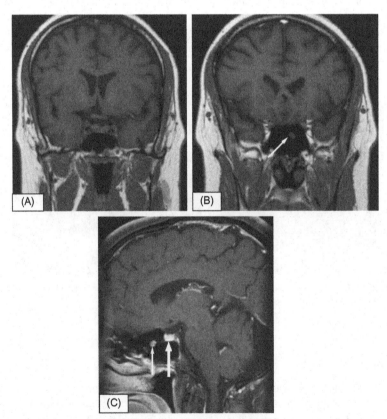

FIGURE 1.17 Ectopic pituitary adenoma: (A) T1-weighted coronal image without contrast, (B) T1-weighted coronal image without contrast, and (C) T1-weighted sagittal image postcontrast.

CYSTIC LESIONS

Rathke Cleft Cyst

The MRI (Figure 1.18) shows the sellar mass with mild displacement of the optic chiasm. The patient had bitemporal visual field deficit and underwent surgery. The sellar mass was confirmed to be a Rathke's cleft cyst on pathology. The images were obtained from a previous publication by our group [1].

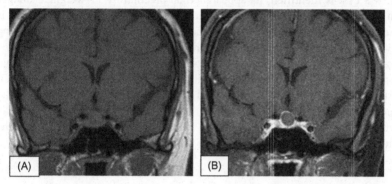

FIGURE 1.18   Rathke cleft cyst: (A) T1-weighted coronal image precontrast, (B) T1-weighted coronal image postcontrast.

(A) The mass is isointense compared to that of a normal-appearing pituitary gland on the precontrast T1-weighted image (i.e., it is not visualized to be different from the surrounding normal pituitary tissue).

(B) The mass does not enhance (appears hypointense) following the administration of contrast.

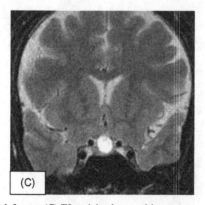

FIGURE 1.18   Rathke cleft cyst: (C) T2-weighted coronal image.

(C) The mass is hyperintense on a T2-weighted image (similar to CSF) suggestive of a cystic lesion.

Pituitary cysts usually have different signal intensities on precontrast images, when compared to those of solid tumors. However, if the cyst is not identified on T1-weighted images, the T2-weighted image will assist the clinician in clarifying the nature of the cyst (Figure 1.18).

Arachnoid Cyst

When a cyst is identified, its appearance on precontrast T1-weighted images can help narrow the differential diagnosis. The signal intensity of a cyst is determined by its fluid content (i.e., the fluid's protein content). Comparing the intensity of CSF in the lateral ventricles to that contained within the cyst can help to differentiate the various cystic lesions. An arachnoid cyst's fluid content has an intensity identical to that of the CSF, whereas a Rathke's cleft cyst may not (depending on the fluid's protein content). Thus, if a cystic lesion has fluid that is not identical to that of the CSF, it is unlikely to be an arachnoid cyst. See Figure 1.19 for an example of an arachnoid cyst and the nature of its fluid content in relation to the CSF.

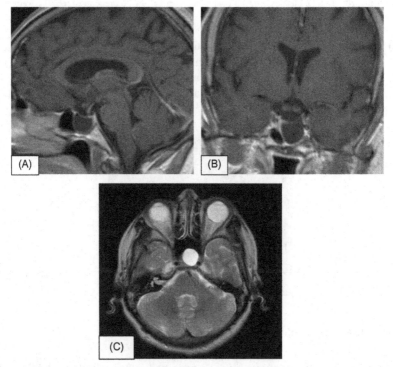

FIGURE 1.19 Arachnoid cyst: (A) T1-weighted sagittal image postcontrast, (B) T1-weighted coronal image postcontrast, and (C) T2-weighted axial image.

Dermoid/Epidermoid Cysts

Dermoid and epidermoid cysts are lined by stratified squamous epithelium. Dermoid cysts, unlike epidermoids, also contain epidermal appendages such as hair follicles, sweat, and sebaceous glands. The sebaceous glands are responsible for the secretion of sebum, which imparts the characteristic appearance of the dermoids on MRI. It is a common misconception that dermoid cysts contain adipose tissue; they do not. Lipocytes originate from the mesoderm, and dermoid cysts (by definition) originate entirely from the ectoderm. If a dermoid cyst did contain adipose tissue it would be termed a teratoma.

Dermoid Cyst

The MRI brain images in Figure 1.20 demonstrate an extra-axial cystic mass with intrinsic T1 hyperintensity on noncontrast imaging (Figure 1.20A, white solid arrow). Note the wisps of T1 hyperintensity in the periphery of the mass on postcontrast imaging (Figure 1.20B, see white dashed arrows), whose etiology is ambiguous since it could either represent the intrinsic hyperintensity indicative of sebum contents, or it could be native enhancement. Thus, it is important to compare precontrast and postcontrast images on the same slice to assess for true enhancement vs. intrinsic hyperintensity.

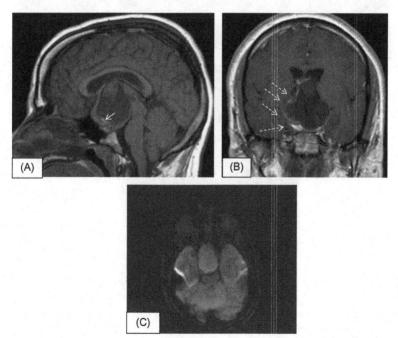

FIGURE 1.20 Dermoid cyst: (A) T1-weighted sagittal image precontrast, (B) T1-weighted coronal image postcontrast, and (C) axial diffusion.

The axial diffusion image (Figure 1.20C) demonstrates that the mass is relatively hyperintense, likely reflecting characteristic diffusion restriction from epithelial–sebaceous components. Note the lack of cystic mass enhancement postcontrast (Figure 1.20B).

Epidermoid Cyst

The epidermoid cyst shows a more uniform character, appearing almost CSF-like, with no frank macroscopic hyperintense signal to suggest sebum, which differentiates a dermoid from an epidermoid cyst. The lesion appears to softly fill the suprasellar cistern, insinuating around structures, with mild mass effect (Figure 1.21).

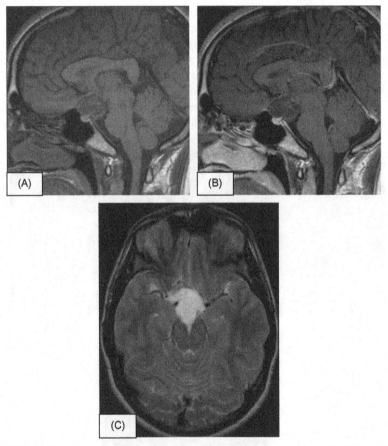

FIGURE 1.21   Epidermoid cyst: (A) T1-weighted sagittal image precontrast, (B) T1-weighted sagittal image postcontrast, and (C) T2-weighted axial image.

PITUICYTOMA

A 61-year-old man presented to the ED complaining of dizziness. CT and subsequently an MRI noted a 1.8 cm sellar mass extending into the suprasellar cistern, elevating and distorting the optic chiasm. The patient did not have any visual field deficits. No cavernous sinus invasion was appreciated. The mass demonstrated fairly homogeneous enhancement after the administration of contrast. At surgery, a vascular, firm, whitish lesion was identified. Histopathologic examination revealed it to be a pituicytoma, a rare brain tumor that is thought to be derived from the parenchymal cells (pituicytes) of the posterior pituitary (Figure 1.22).

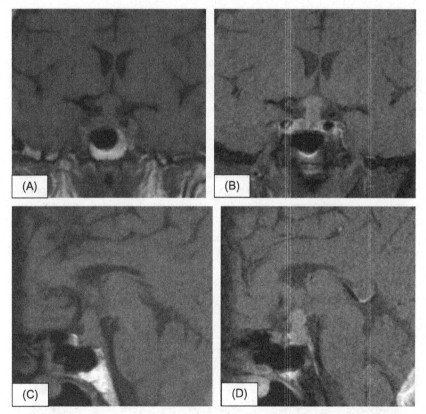

FIGURE 1.22   Pituicytoma: (A) T1-weighted coronal image precontrast, (B) T1-weighted coronal image postcontrast, (C) T1-weighted sagittal image precontrast, (D) T1-weighted sagittal image postcontrast.

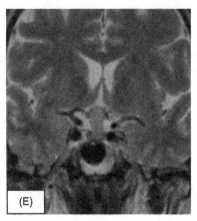

FIGURE 1.22   Pituicytoma: (E) T2-weighted coronal image.

CHORDOMA

A 39-year-old woman presented with headaches. A CT scan was performed and a sellar mass was noted. Subsequently an MRI revealed a 3.0 cm × 3.4 cm × 3.3 cm mass, in an expanded sella, that extended into the suprasellar cistern. The mass was slightly hypointense to gray matter on T1-weighted images, and showed mild heterogeneous enhancement after the administration of contrast. This mass elevated and distorted the right half the optic chiasm (Figure 1.23B, white arrow). It also caused leftward deviation of the pituitary. In the T1-weighted coronal precontrast image (Figure 1.23A), the dashed white line identifies the neurohypophysis (posterior pituitary gland), which appears hyperintense on precontrast images (the anterior pituitary gland does not typically appear hyperintense prior to the administration of contrast). Note the area of normal pituitary tissue displaced to the left, observed at the site of stalk termination (black arrow), on the postcontrast T1-weighted coronal image.

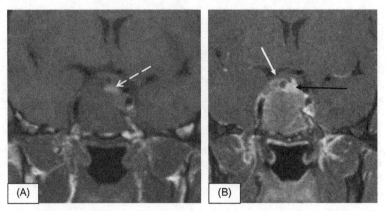

FIGURE 1.23   Chordoma: (A) T1-weighted coronal image precontrast, (B) T1-weighted coronal image postcontrast.

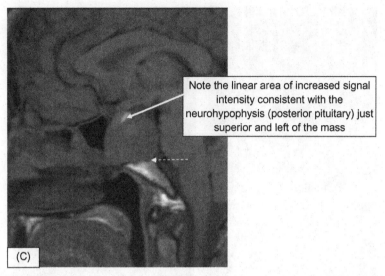

Note the linear area of increased signal intensity consistent with the neurohypophysis (posterior pituitary) just superior and left of the mass

(C)

FIGURE 1.23 Chordoma: (C) T1-weighted sagittal image precontrast (postcontrast sagittal image is unavailable).

Surgical pathology was consistent with a chordoma, a tumor that arises from cellular remnants of the notochord. One of the most common locations for occurrence is within the clivus, and thus they may frequently involve the sella. White dashed arrow (Figure 1.23C): the clivus adjacent to this mass demonstrates an irregular and thin hypointense cortical rim, suggestive of erosion, which is superimposed on more general remodeling. The marrow in the remainder of the clivus is normal in signal intensity.

EMPTY SELLA (FIGURE 1.24)

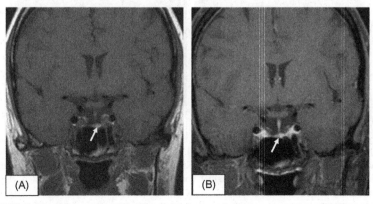

(A) (B)

FIGURE 1.24 Empty sella: (A) T1-weighted coronal image precontrast, (B) T1-weighted coronal image postcontrast.

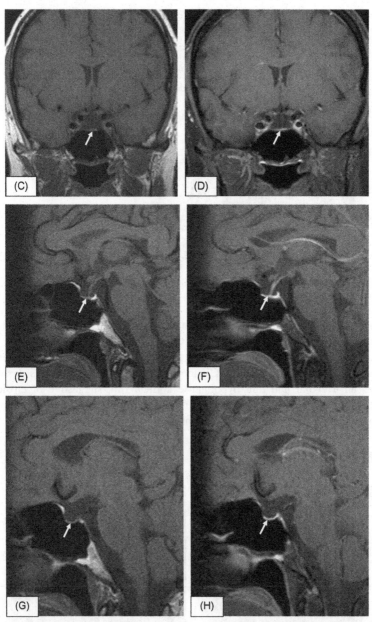

FIGURE 1.24 Empty Sella: (C) T1-weighted coronal image precontrast, (D) T1-weighted coronal image postcontrast, (E) T1-weighted sagittal image precontrast, (F) T1-weighted sagittal image postcontrast, (G) T1-weighted sagittal image precontrast, and (H) T1-weighted sagittal image postcontrast.

Note that the pituitary stalk terminates into an empty sella, which is shallow, and only a small amount of pituitary tissue is identifiable at the base of the sella (white arrow). Of the figures included, 1.24B, 1.24D, 1.24F, and 1.24H were all previously published by our group [1]. An empty sella is often an incidental radiological finding with no evidence of any hormonal deficiency, although it can be associated with variable degrees of hypopituitarism. It may also develop following pituitary surgery or radiotherapy.

Although the term empty sella is used, the sella is not actually empty: in general, a pituitary gland height of 2 mm or less is referred to as an empty sella, whereas if the height of the gland is 3−4 mm it is referred to as a partial empty sella.

Partial Empty Sella (Figure 1.25)

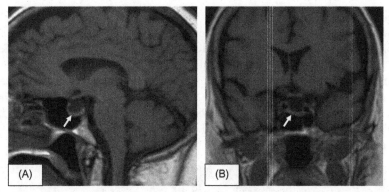

FIGURE 1.25 Partial empty sella: (A) T1-weighted sagittal image precontrast, and (B) T1-weighted coronal image precontrast. Note the small 2−3 mm rim of pituitary tissue at the base of the sella (white arrow).

GERMINOMA

A 33-year-old man presented with a recent onset of seizures. Evaluation led to the discovery of a large sellar mass. The mass, with its epicenter in the sella, demonstrated suprasellar extension, with displacement and distortion of the optic chiasm. The mass also invaded the right cavernous sinus (white dashed arrow) with partial encirclement of the internal carotid artery, and possibly invasion of the left cavernous sinus. It measured 2.2 cm × 2.9 cm × 1.3 cm at its greatest CC, transverse, and AP dimensions, respectively. The lesion was bi-lobed (Figure 1.26A), commonly referred to as a "snowman configuration." In addition, there was an abnormal, predominantly cystic signal within the right thalamus/inferior basal ganglia (white arrow, Figure 1.26B), measuring approximately 1.5 cm × 1.8 cm × 2.2 cm at its greatest CC, transverse, and AP dimensions, respectively, which represents an extension of the sellar mass (a direct connection between the two lesions could not be identified within a single image). The mass exhibits mild enhancement.

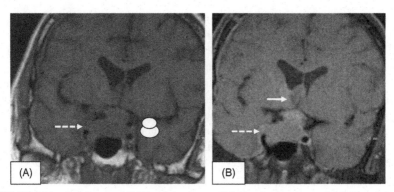

FIGURE 1.26   Germinoma: (A) T1-weighted coronal image precontrast, and (B) T1-weighted coronal image postcontrast.

CRANIOPHARYNGIOMA

Craniopharyngiomas are rare tumors that arise from remnants of Rathke's pouch, the embryological origin of the anterior pituitary gland. These lesions typically involve the sella and extend superiorly. They may be solid, cystic, or, as is most common, a combination. Calcification of these tumors is common. Accordingly, when a mixed solid/cystic tumor containing calcification is noted on MRI, one should strongly consider craniopharyngioma as a likely diagnosis. There is a bimodal age distribution, with one peak in children between 5 and 14 years, and a second peak in adults in the fifth to seventh decades of life [9].

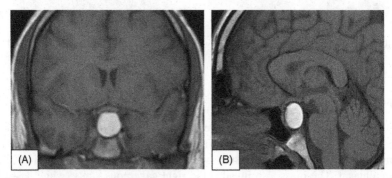

FIGURE 1.27   Craniopharyngioma: (A) T1-weighted coronal image precontrast, (B) T1-weighted sagittal image precontrast.

Note the hyperintense appearance of the lesion on T1-weighted imaging prior to administration of IV contrast (Figure 1.27A and B), a finding that is characteristic of a craniopharyngioma, and likely due to a high concentration of cholesterol crystals.

On T2-weighted imaging (Figure 1.27C), the sellar mass is iso- to hypointense compared to the brain parenchyma. This is not suggestive of an underlying hemorrhagic pituitary lesion, since blood usually appears white on T2-weighted images during subacute hemorrhage (hours to months after

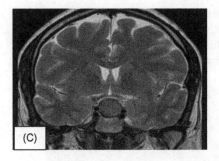

FIGURE 1.27 Craniopharyngioma: (C) T2-weighted coronal image.

an event). This is important to recognize, as subacute apoplexy/hemorrhage can also have a hyperintense appearance in T1-weighted imaging (see later text on pituitary apoplexy).

The histologic findings of squamous epithelium with no granular layer, debris rich in cholesterol crystals, focal calcifications, and minute fragments of fibrous tissue, together with the reported radiologic findings, are consistent with a craniopharyngioma.

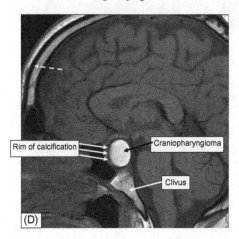

FIGURE 1.27 Craniopharyngioma: (D) T1-weighted sagittal image without contrast.

Note the black rim at the anterior aspect of the mass in Figure 1.27D. This is a rim of calcification. Highly calcified structures, like cortical bone, which contains no fat and few mobile protons, appear black on MRI (see white dashed arrow pointing to a black line representing the outer table of the calvarium, with a subjacent bright layer representing fat within the calvarial marrow, which is adjacent to another black line representing the inner table of the calvarium).

MENINGIOMA

Meningiomas arise from the meninges, the membranous layers surrounding the central nervous system (CNS). They are usually benign, and can arise anywhere in the CNS, including near the sella.

Note the avid and uniform enhancement of the sellar mass post-IV contrast (thick white arrow), a feature that is characteristic of a meningioma and one that may assist in differentiating it from that of a pituitary adenoma, which does not usually enhance to the same degree. The enhancement and thickening of the adjacent dura is referred to as "dural tail" and is a characteristic for an underlying meningioma or other dural-based process (Figure 1.28A–D).

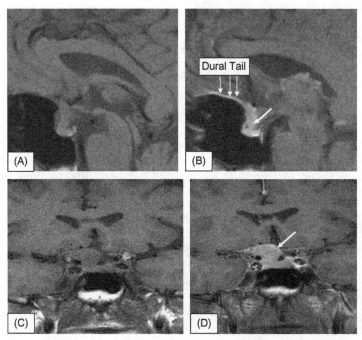

FIGURE 1.28   Meningioma: (A) T1-weighted sagittal image precontrast, (B) T1-weighted sagittal image postcontrast, (C) T1-weighted coronal image precontrast, (D) T1-weighted coronal image postcontrast.

Another patient (Figure 1.28) with a meningioma demonstrates again a uniform avid enhancement on T1-weighted imaging (white arrow).

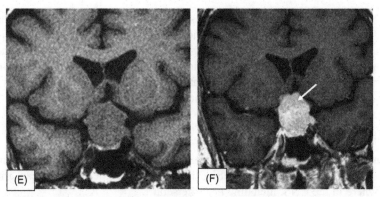

FIGURE 1.28   (E) T1-weighted coronal image precontrast, and (F) T1-weighted coronal image postcontrast.

PITUITARY HYPERPLASIA

Pituitary hyperplasia may be a source of confusion for the clinician and may lead to inappropriate referral for surgery. A rather specific finding for pituitary hyperplasia is the "hilltop" or "mountain peak" sign (figure 1.29A). It is often seen in the context of longstanding untreated primary hypothyroidism (figure 1.29B), menopause (figure 1.29C—E), and pregnancy, as well as occasionally in adolescent and young women. Figure 1.29C—E demonstrates pituitary hyperplasia in a 51-year-old postmenopausal woman with FSH 106 mU/mL and LH 75.7 mU/mL.

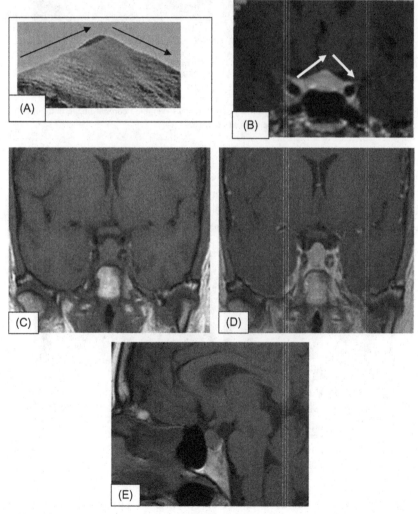

FIGURE 1.29 Pituitary hyperplasia: (A) hilltop (note the incline-peak-down slope, from left to right; see arrows), (B) T1-weighted coronal image postcontrast, (C) T1-weighted coronal image precontrast, (D) T1-weighted coronal image postcontrast, and (E) T1-weighted sagittal image precontrast. The coronal images were obtained from a previous publication by our group [1].

ANATOMIC VARIATIONS

It is important for clinicians to recognize pituitary/sellar anatomical variations in order to avoid unnecessary referral and work-up.

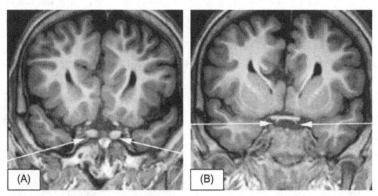

FIGURE 1.30 Duplication of the pituitary gland: (A) T1-weighted coronal image without contrast, (B) T1-weighted coronal image without contrast.

A patient with two pituitary glands (see arrows) and two pituitary stalks [see arrows in (B)], obtained from Manjila et al. [10]. Duplication of the pituitary gland is an extremely rare malformation. The patient is an 11-year-old girl with multiple craniofacial anomalies and severe intellectual disability.

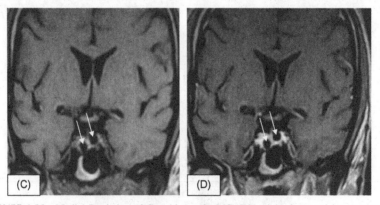

FIGURE 1.30 Medial Deviation of Carotid Arteries: (C) T1-weighted coronal image precontrast, (D) T1-weighted coronal image postcontrast.

Patient with medial deviation of carotid arteries (arrows) (aka "kissing carotids"). The images were obtained from a previous publication by our group [1].

(E-H) Patient with ectopic bright spot/ectopic posterior pituitary.

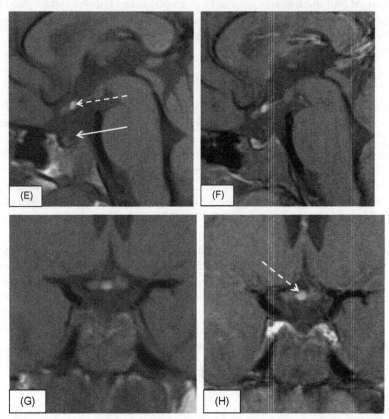

FIGURE 1.30   Ectopic Bright Spot/Ectopic Posterior Pituitary: (E) T1-weighted sagittal image precontrast, (F) T1-weighted sagittal image postcontrast, (G) T1-coronal image precontrast, and (H) T1-coronal image postcontrast.

The pituitary bright spot is usually observed within the sella turcica in the posterior pituitary. On T1-weighted images, the anterior pituitary typically appears isointense, whereas the posterior pituitary appears hyperintense. Although the exact etiology of the bright spot is unknown, it is thought to be secondary to the high concentration of neurosecretory granules contained within the posterior pituitary. Almost no patients with central diabetes insipidus have a posterior bright spot. At the same time, about 10% of subjects without an underlying hypothalamic–pituitary disorder do not have a bright spot [11]. The lack of the normal posterior pituitary bright spot, in the absence of central diabetes insipidus, should encourage a search for an ectopic posterior pituitary that has failed to migrate inferiorly from the hypothalamus during embryologic development. In summary, while the absence of a pituitary bright spot is non-specific and may be normal, visualization of an ectopic pituitary bright spot is abnormal. In the accompanying figures, note the position of the posterior pituitary bright spot immediately posterior to the optic chiasm (dashed white arrow), rather than in its usual position in the posterior aspect of the sella (white arrow), immediately posterior to the anterior pituitary (Figure 1.30A).

PITUITARY STALK TRANSECTION

Pituitary stalk transection was described after the introduction of MRI and comprises a small anterior pituitary gland, thin or absent infundibulum after gadolinium administration, and an ectopic location of the posterior pituitary [12].

The accompanying images (Figures 1.31A–D) were obtained from a 36-year-old woman who presented to the adult endocrinologist with multiple anterior pituitary hormone deficiencies. GH deficiency was diagnosed after she stopped growing around the age of 5. GH therapy was initiated and was continued until the patient was 18 years of age. She subsequently developed hypothyroidism in her early 20s, requiring progressively higher doses of levothyroxine. In addition, she did not develop secondary sexual characteristics or have menstrual cycles until she was started on oral contraceptives in her early 20s. She was diagnosed with polyglandular autoimmune deficiency in her early 30s and was started on dexamethasone 0.5 mg at bedtime. This was subsequently changed to therapy with hydrocortisone, 10 mg in morning and 5 mg in the afternoon (an appropriate daily dose based on her body habitus).

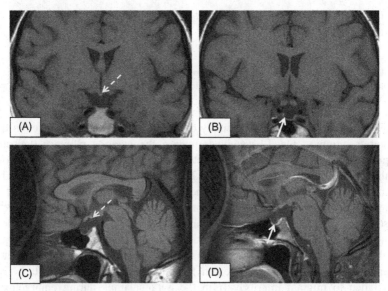

FIGURE 1.31 Pituitary stalk transection: (A) T1-weighted coronal image precontrast, (B) T1-weighted coronal image precontrast, (C) T1-weighted sagittal image precontrast, and (D) T1-weighted sagittal image postcontrast.

The T1-weighted coronal image precontrast demonstrated an ectopic posterior pituitary along the floor of the hypothalamus (white dashed line). The pituitary stalk was very small and demonstrated only minimal enhancement following the administration of contrast. A small rim of anterior pituitary tissue was noted along the floor of the sella (solid white arrow). This clinical case and some of the images were previously published by Ioachimescu et al. [13] (Figure 1.31).

VASCULAR LESIONS (ANEURYSM)

Identifying vascular lesions involving the sella is important. Additional imaging studies (MR angiogram, CT angiogram, etc.) are often required to further evaluate lesions that have MRI characteristics suggesting the possibility of a vascular lesion. Imaging features that are suggestive of a vascular etiology are artifacts due to flowing blood or a blood clot. When flowing protons have a speed above a certain threshold (which depends on the sequence), they no

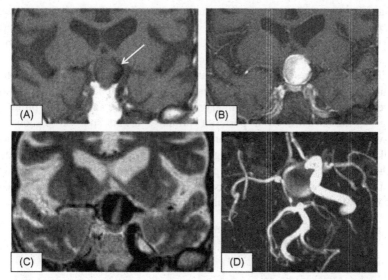

FIGURE 1.32 Vascular lesions (aneurysm): (A) T1-weighted coronal image precontrast of a large sellar aneurysm, showing a flow void (see white arrow), greatest on the left, with other heterogeneous signal likely related to turbulent blood flow. (B) T1-weighted coronal image postcontrast, showing a smooth gradation of enhancement across the aneurysm characteristic of flowing blood with turbulence. (C) T2-weighted coronal image, showing marked loss of signal representing a large flow void due to moving blood. The central vertical focus is a possible web or septation through the central aspect of the aneurysm. (D) Finally, aneurysms can have a mixed appearance on time-of-flight MRA, with the image showing reduced flow-related enhancement due to a combination of slow and turbulent flow within the aneurysm, which causes relative signal loss compared to the fast laminar flow in the adjacent internal carotid arteries.

longer contribute any MRI signal and appear black on an image, often termed a "flow void." This feature is particularly apparent on a T2-weighted sequence, and less so on a noncontrast T1-weighted sequence. After the administration of contrast there is often avid enhancement, but the pattern of enhancement may have a smooth gradation of intensity across the lesion (sometimes termed "shading"), whose pattern does not conform to any expected underlying anatomy. Finally, the presence of clot will create a heterogeneous signal adjacent to flow voids, which does not enhance. Sometimes the clot will have a web-like or crecentic morphology (Figure 1.32).

INFILTRATIVE DISORDERS

Lymphocytic Hypophysitis

Lymphocytic hypophysitis is an uncommon pituitary disorder that predominantly affects young females. When it does occur, there is a tendency to present in the early postpartum period. Typical symptoms and signs include headaches, visual field deficits, pituitary dysfunction, etc. It is characterized by autoimmune inflammation (lymphocytic invasion) of the pituitary gland. MRI of the pituitary region usually demonstrates an enhancing soft tissue mass involving the anterior and posterior pituitary as well as the infundibulum, which is often markedly thickened.

The patient presented with DI, and no obvious sellar mass was noted on imaging (Figure 1.33A). However, the presence of DI, which is almost never seen in patients with pituitary adenomas before undergoing pituitary surgery, the absence of a posterior pituitary bright spot on precontrast T1-weighted sagittal image (Figure 1.33B), and the moderately thickened stalk (Figure 1.33C) were suggestive of an underlying inflammatory pathology. The hyperintense area posterior to the pituitary gland on precontrast sagittal view (Figure 1.33B) is the posterior wall of the sella and should not be confused with posterior pituitary.

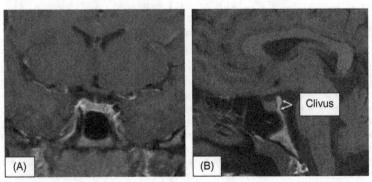

FIGURE 1.33 Lymphocytic hypophysitis: (A) T1-weighted coronal image postcontrast, (B) T1-weighted sagittal image precontrast.

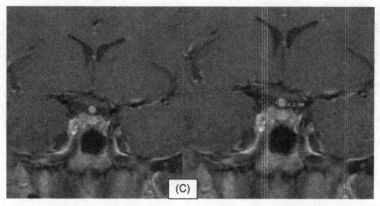

FIGURE 1.33   Lymphocytic hypophysitis: (C) T1-weighted coronal image postcontrast.

The patient was followed, and because of further stalk thickening upon reassessment (Figure 1.33D and E), underwent a biopsy, which confirmed lymphocytic hypophysitis.

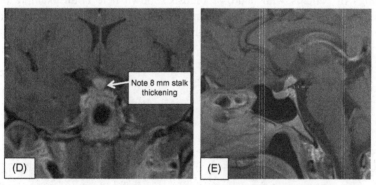

FIGURE 1.33   Lymphocytic hypophysitis: (D) T1-weighted coronal image postcontrast, and (E) T1-weighted sagittal image postcontrast.

Langerhans Cell Histiocytosis

Langerhans cell histiocytosis is a rare disorder characterized by the abnormal proliferation of histiocytes; it can affect any organ, including the pituitary gland and stalk. On T1-weighted imaging the lesions typically enhance after the administration of IV contrast. In the hypothalamic—pituitary region, infundibular involvement, with thickening and enhancement seen on MRI (see white arrows, Figure 1.34), has been reported to occur in 50% of subjects (which explains the high frequency of diabetes insipidus upon presentation); pronounced hypothalamic mass lesions in 10%; and infundibular atrophy in 29% [14]. The presence of additional pathology in other organs/locations, particularly in bone, should raise suspicion for this systemic disorder (see Figure 1.34C). In the example, note the abnormal thickening and enhancement of the pituitary stalk, which appears contiguous with tissue of similar character in the posterior sella.

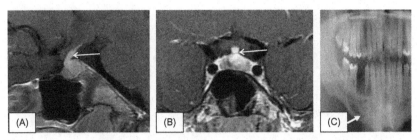

FIGURE 1.34 Langerhans cell histiocytosis: (A) T1-weighted sagittal image postcontrast, (B) T1-weighted coronal image postcontrast, (C) a well-circumscribed lytic lesion with no evident cortex destruction is depicted at the right part of the mandible's body. Image obtained from Anastasilakis et al. [15].

Neurosarcoidosis

Patients with neurosarcoidosis can present with variable imaging findings and clinical manifestations. Stalk thickening with uniform enhancement, along with leptomeningeal enhancement, are among the most common MR imaging findings in patients with neurosarcoidosis. Shown here is an example of a patient with neurosarcoidosis and panhypopituitarism with a thickened stalk (>4 mm) that enhances after administration of IV contrast (Figure 1.35).

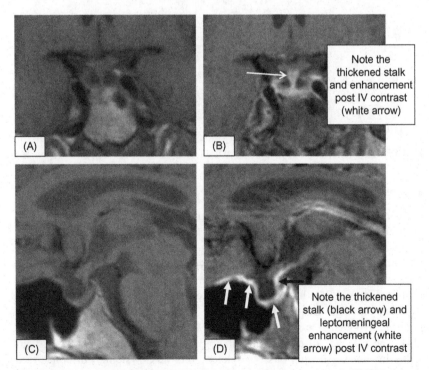

Note the thickened stalk and enhancement post IV contrast (white arrow)

Note the thickened stalk (black arrow) and leptomeningeal enhancement (white arrow) post IV contrast

FIGURE 1.35 Neurosarcoidosis: (A) T1-weighted coronal image precontrast, (B) T1-weighted coronal image postcontrast, (C) T1-weighted sagittal image precontrast, and (D) T1-weighted sagittal image postcontrast.

Wegener's Granulomatosis

Wegener's granulomatosis is a form of vasculitis that affects small- and medium-size vessels in many organs. The upper respiratory tract, lungs, and kidneys are often involved, but it can affect any organ system, though rarely involving the pituitary. Isolated pituitary involvement is extremely rare, but has been reported [16].

A 37-year-old woman with Wegener's granulomatosis presented with an acute onset of left third nerve palsy and diplopia. Pituitary MRI noted a sellar mass with suprasellar extension measuring 1.9 cm × 1.4 cm × 2.3 cm. The mass was isointense compared to brain gray matter on precontrast images, and showed diffuse heterogeneous enhancement. A mass effect was manifest as elevation and distortion of the optic chiasm. No intrinsic T1 hyperintensity was noted to suggest recent hemorrhage. Pathology was consistent with Wegener's granulomatosis (Figure 1.36).

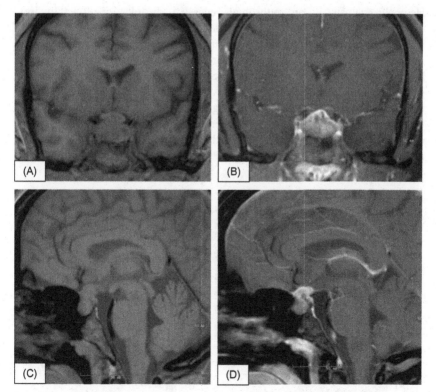

FIGURE 1.36 Wegener's granulomatosis: (A) T1-weighted coronal image precontrast, (B) T1-weighted coronal image postcontrast, (C) T1-weighted sagittal image precontrast, (D) T1-weighted sagittal image postcontrast.

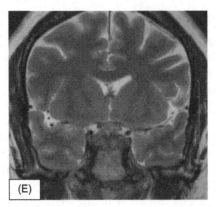

FIGURE 1.36   Wegener's Granulomatosis: (E) T2-weighted coronal image.

Hemochromatosis

Hemochromatosis is an autosomal recessive disorder resulting in iron overload. As a result, hemosiderin collects throughout the body, sometimes including the pituitary. An imaging abnormality characteristic of hemochromatosis (or other conditions that result in hemosiderin deposition, i.e., chronic hemorrhagic products) is the signal loss or "blackness" seen throughout the gland on the T2 sequences (see white arrow, Figure 1.37D). This is caused by microscopic dephasing of the magnetic field by hemosiderin. Similar blackness is not seen on the T1 sequences (either the pre- or postcontrast) due to the vastly different values of the repetition time (TR) used, which make T2 sequences more sensitive to dephasing from hemosiderin.

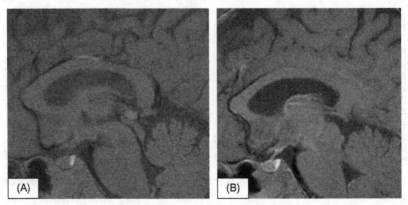

FIGURE 1.37   Hemochromatosis: (A) T1-weighted sagittal image precontrast, (B) T1-weighted sagittal image postcontrast.

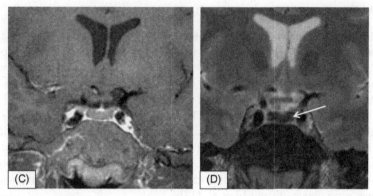

FIGURE 1.37 Hemochromatosis: (C) T1-weighted coronal image postcontrast, and (D) T2-weighted coronal image.

METASTASES

Patients with pituitary metastases usually have a history of a primary malignancy. A biopsy may be required to determine the pathology. Metastatic lesions involving the pituitary should be a concern in patients with an atypical-appearing sellar mass with a history of malignancy (especially breast, lung, and renal cell cancers). The lesions are typically heterogeneously enhancing and have an atypical or complex morphology. Shown here are some selected images from a 56-year-old woman with pituitary metastasis confirmed via biopsy. She had a history of metastatic breast cancer. Note the heterogeneous enhancement and stalk thickening (>4 mm; see white arrow), findings similar to those seen with infiltrative pituitary lesions (Figure 1.38).

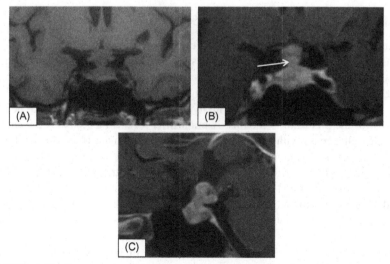

FIGURE 1.38 Pituitary metastasis: (A) T1-weighted coronal image precontrast, (B) T1-weighted coronal image postcontrast, and (C) T1-weighted sagittal image postcontrast.

PRIMARY PITUITARY LYMPHOMA

Primary CNS lymphomas are rare, and when they do occur, they usually occur in immunocompromised individuals. Primary pituitary lymphoma, particularly in an immunocompetent individual, is an exceedingly rare occurrence [17].

A 41-year-old woman presented with polyuria and polydipsia, weight loss, galactorrhea, and amenorrhea. The patient was found to have an elevated serum PRL, as well as DI and hypopituitarism. Postcontrast sagittal and coronal MR images (Figure 1.39A and B) showed a sellar mass with thickening of the pituitary stalk and avid homogenous enhancement (white arrow). Pathology was consistent with a high grade, large B cell lymphoma. Images were obtained from Li et al. [17].

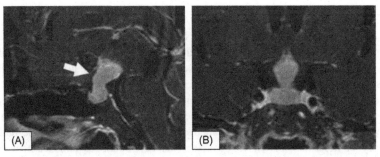

FIGURE 1.39 Primary pituitary lymphoma: (A) T1-weighted sagittal image postcontrast, and (B) T1-weighted coronal image postcontrast.

PITUITARY INFECTION/ABSCESS

Pituitary abscess are rare, but can be very debilitating. Patients often present with complaints and physical findings of a sellar mass (headache, visual changes, endocrine abnormalities) rather than that of an infection, with only one-third of patients typically presenting with fever, leukocytosis, etc., [18].

A 36-year-old man presented with hypopituitarism and a complaint of headaches for the preceding 3 months. A brain MRI revealed a 1.4 cm × 2.0 cm × 1.6 cm sellar mass with a peripheral ring of enhancement (Figure 1.40, long black arrow) surrounding the hypointense sellar mass. Stalk thickening was also observed (short arrow). Throughout the body, an imaging characteristic of an abscess is a rounded nonenhancing lesion with a ring-enhancing periphery and adjacent inflammatory reaction, and this feature also pertains to the pituitary gland. The primary differential is that of a cyst, but these usually have a thin, smooth enhancing rim with smooth contours, and without any adjacent inflammatory changes. The case description and images were obtained from Walia et al. [19].

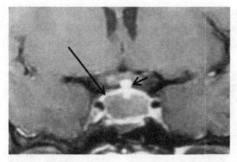

FIGURE 1.40 Pituitary abscess: T1-weighted coronal image postcontrast.

POSTOPERATIVE PITUITARY IMAGING

Visible Fat Pad

This case involves a patient with status post-pituitary surgery for a macro-adenoma with subsequent partial empty sella and a visible fat pad at the floor of the sella. Some fat pads that are placed at the completion of trans-sphenoidal resection to close the surgically-made access can vascularize and become permanently visualized in the area of the sella. The fat pad is visible as a hyperintense (white) area at the base of sella (see arrows) (Figure 1.41).

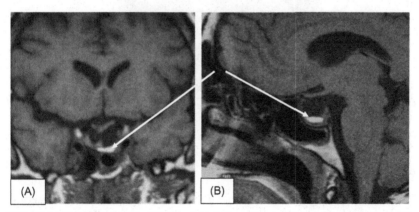

FIGURE 1.41 Visible fat pad: (A) T1-weighted coronal image without contrast, and (B) T1-weighted sagittal image without contrast.

Sellar Remodeling Post-Pituitary Surgery

An empty sella may be an incidental radiological finding, typically with no pituitary hormone deficiencies, or it may develop following pituitary surgery or radiotherapy. The accompanying images (Figure 1.42A–C) are from a patient with a nonfunctional pituitary macroadenoma after surgery and radiotherapy. Pituitary MRI postsurgery is preferably delayed for 1−4 months or more (usually about 3 months) to avoid acute postsurgical changes (edema and mass effect) and to allow scar formation to stabilize.

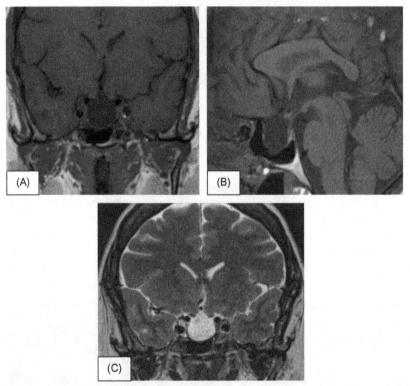

FIGURE 1.42 Empty sella with sellar remodeling: (A) T1-weighted coronal image precontrast, (B) T1-weighted sagittal image precontrast, and (C) T2-weighted coronal image.

APOPLEXY

Hemorrhage

Pituitary hemorrhage can occur in a variety of settings. It has been observed in patients taking anticoagulants, as a complication of head trauma, and in patients with underlying pituitary pathology (macroadenoma, etc.). However, it can also occur in patients without identifiable risk factors for hemorrhage, and in those without any underlying pituitary pathology. The appearance of apoplexy on MRI can vary considerably, and is strongly affected by the time interval preceding the MRI and the characteristic temporal evolution of blood products.

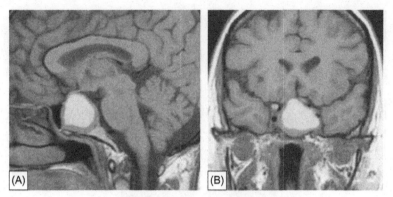

FIGURE 1.43 Pituitary apoplexy: (A) T1-weighted sagittal image precontrast, (B) T1-weighted coronal image precontrast.

A hyperintense lesion on T1-weighted imaging, without IV contrast (Figure 1.43), is most concerning for a hemorrhage (specifically representing methemoglobin); however, this appearance can also be seen from a cystic mass containing proteinacious material, such as craniopharyngioma. Astute clinicians may use their knowledge of the appearance of blood on T1- and T2-weighted images, and how that appearance varies with time, to differentiate hemorrhage from other T1 hyperintensities.

While the time evolution of blood products on various sequences is complicated, some mnemonics can be useful to memorize the pattern: on T1-weighted imaging, without contrast, in the acute setting (within first couple of hours), hemorrhage often appears "Gray," in the subacute setting (hours to months) it often appears "White," and in the chronic setting (months to years) it usually appears "Black." This can be memorized with the mnemonic: **G**eorge **W**ashington **B**ridge.

On T2-weighted imaging, in the acute setting (hours), blood often appears "Black," in the subacute setting (hours to months) it often appears "White," and in the chronic setting (months to years) it usually appears "Black." This can be memorized with the mnemonic: Oreo® cookie: **B**lack/**W**hite/**B**lack (cookie, cream, then cookie; i.e., the colors of an Oreo® cookie).

Note that the periphery of the lesion in question in Figure 1.44A−C appears hyperintense on both the T1- and T2-weighted images, a finding suggestive of subacute hemorrhage. The center appears more hypointense, and is smoothly graduated from the periphery. This observation highlights the fact that the speed of evolving hemosiderin oxidation depends on the local tissue oxygenation tension. Thus, the center of a hematoma retains its

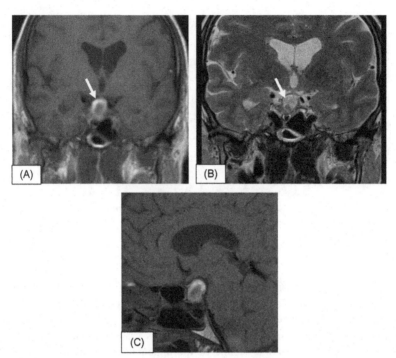

FIGURE 1.44 Pituitary apoplexy: (A) T1-weighted coronal image without contrast, (B) T2-weighted coronal image, (C) T1-weighted sagittal image without contrast in the same patient with apoplexy.

fresh blood character longer due to low oxygen perfusion and tension, whereas the periphery oxidizes more rapidly due to its proximity to higher oxygen perfusion and tension.

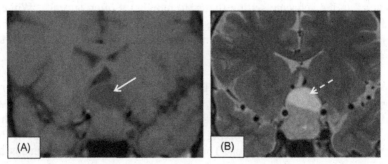

FIGURE 1.45 Cystic pituitary macroadenoma: (A) T1-weighted coronal image without contrast, and (B) T2-weighted coronal image.

Note that in the accompanying images (Figure 1.45A and B), in contrast to the previous imaging findings with hemorrhage (Figure 1.44A and B), the cystic pituitary macroadenoma has a hypointense (black) appearance on T1-weighted imaging (white arrow), but hyperintense (bright white) on T2-weighted imaging (white dashed arrow). This appearance on T1- and T2-weighted imaging is suggestive of a cystic lesion, not hemorrhage.

Infarction (Non-Hemorrhagic)

Perhaps the best example of non-hemorrhagic pituitary infarction is Sheehan syndrome, or postpartum pituitary infarction, which can occur as a result of excessive blood loss and hypotension that can complicate deliveries (Figure 1.46).

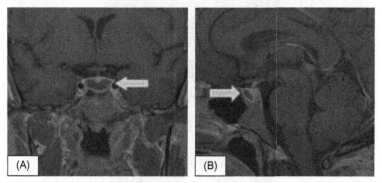

FIGURE 1.46 Non-hemorrhagic pituitary infarction: (A) T1-weighted coronal image postcontrast, (B) T1-weighted sagittal image postcontrast.

In Figures 1.46A–B, note the enlarged pituitary gland with rim of enhancement and a central area that lacks enhancement. Images were obtained from Morani et al. [20].

The next set of images, Figures 1.46C–D, were obtained 1 year postinfarction. Note the resultant atrophy of the anterior pituitary gland. Images were also obtained from Morani et al. [20].

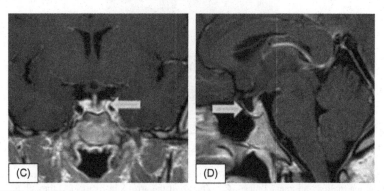

FIGURE 1.46 Non-hemorrhagic pituitary infarction, 1 year post infarction: (C) T1-weighted coronal image postcontrast, and (D) T1-weighted sagittal image postcontrast.

VOLUME AVERAGING

Volume averaging is an imaging artifact that occurs when the object of interest (the pituitary gland, for example) is only partially within the slice (volume) of interest. When this occurs, the resulting pixel value at the location of interest is the average of both the object and its immediate surroundings (e.g., air in the sphenoid sinus or CSF in the front of the brainstem). An imaging artifact from volume averaging may be interpreted as a pituitary adenoma (Figure 1.47).

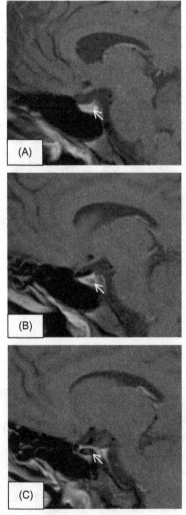

FIGURE 1.47   Volume averaging: Three consecutive cuts of T1-weighted sagittal images post-contrast. The hypodensity that is suggestive of a possible adenoma in the posterior aspect of the pituitary in image slice (B) (see white arrow) is an "average" of the pituitary tissue [image (A)] and the internal carotid artery [image (C)]. The tip of the arrows in figures (A) through (C) are pointing to the same location.

CT SELLA

Although MRI is the preferred imaging modality to evaluate sellar pathology/anatomy, CT may be considered when the cost of MR imaging is an issue, or when one needs to rule out a large sellar mass associated with mild hyperprolactinemia secondary to stalk effect. This may include patients in whom a medication associated with hyperprolactinemia cannot be safely withheld (such as someone with psychosis on antipsychotic medications). Additional situations where CT may be indicated would be in patients with a pacemaker, MRI-incompatible metal pieces or implants, or shrapnel. CT may also be helpful to give clues about the presence of an underlying craniopharyngioma by the demonstration of calcification. Finally, the superior osseous visualization of CT provides optimal imaging for presurgical planning of a trans-sphenoidal approach for resection (Figures 1.48–1.51).

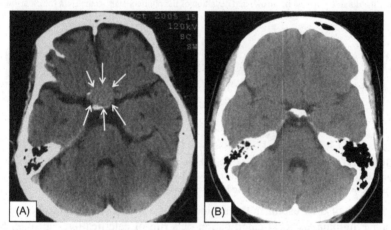

FIGURE 1.48 (A) Transverse CT brain without IV contrast obtained after a motor vehicle accident demonstrated a sellar mass (see arrows), and (B) normal transverse CT sellar image for comparison. Image (A) was obtained from a previous publication by our group [1].

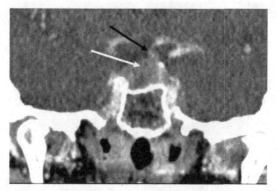

FIGURE 1.49 Coronal image of a patient with sellar mass on CT. The pituitary stalk (black arrow) can be seen deviated to the left side with no displacement on the optic chiasm. The white arrow highlights the sellar mass.

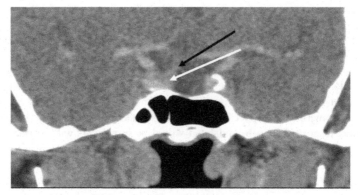

FIGURE 1.50 CT of a patient with pituitary macroadenoma after surgery. The patient underwent CT instead of MRI since he had a pacemaker. Note the pituitary gland along the right lateral margin of the dorsum sella (white arrow) and deviation of the pituitary stalk to the right (black arrow).

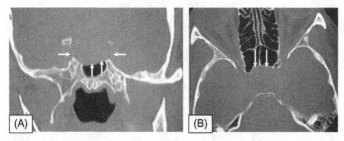

FIGURE 1.51 (A) CT, coronal image with bone windowing, and (B) CT, axial image with bone windowing, of a patient with a sellar mass which shows expansion and remodeling of the sella, whose contours suggest a large isodense lesion (see white arrows).

CT can also be useful in the evaluation of a lesion suspected to be a craniopharyngioma, as focal calcification or a rim of calcification can often be visualized better on CT images [see white arrows in both brain (left) and bone (right) windows in Figure 1.52].

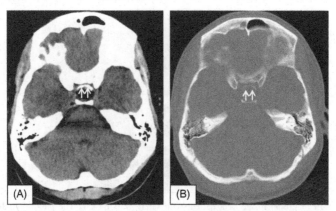

FIGURE 1.52 Craniopharyngioma: (A) CT, axial image, brain window, and (B) CT, axial, bone window.

REFERENCES

[1]   Orija IB, Weil RJ, Hamrahian AH. Pituitary incidentaloma. Best Pract Res Clin Endocrinol Metab 2012;26(1):47–68.

[2]   Simmons GE, Suchnicki JE, Rak KM, Damiano TR. MR imaging of the pituitary stalk: size, shape, and enhancement pattern. AJR Am J Roentgenol 1992;159(2):375–7.

[3]   Ahmadi H, Larsson EM, Jinkins JR. Normal pituitary gland: coronal MR imaging of infundibular tilt. Radiology 1990;177(2):389–92.

[4]   Tsunoda A, Okuda O, Sato K. MR height of the pituitary gland as a function of age and sex: especially physiological hypertrophy in adolescence and in climacterium. AJNR Am J Neuroradiol 1997;18(3):551–4.

[5]   Wolpert SM, Molitch ME, Goldman JA, Wood JB. Size, shape, and appearance of the normal female pituitary gland. AJR Am J Roentgenol 1984;143(2):377–81.

[6]   Elster AD, Sanders TG, Vines FS, Chen MY. Size and shape of the pituitary gland during pregnancy and post partum: measurement with MR imaging. Radiology 1991;181 (2):531–5.

[7]   Fleseriu M, Lee M, Pineyro MM, Skugor M, Reddy SK, Siraj ES, et al. Giant invasive pituitary prolactinoma with falsely low serum prolactin: the significance of "hook effect." J Neurooncol 2006;79(1):41–3.

[8]   Zerikly RK, Eray E, Faiman C, Prayson R, Lorenz RR, Weil RJ, et al. Cyclic Cushing syndrome due to an ectopic pituitary adenoma. Nat Clin Pract Endocrinol Metab 2009;5 (3):174–9.

[9]   Bunin GR, Surawicz TS, Witman PA, Preston-Martin S, Davis F, Bruner JM. The descriptive epidemiology of craniopharyngioma. J Neurosurg 1998;89(4):547–51.

[10]   Manjila S, Miller EA, Vadera S, Goel RK, Khan FR, Crowe C, et al. Duplication of the pituitary gland associated with multiple blastogenesis defects: duplication of the pituitary gland (DPG)-plus syndrome. Case report and review of literature. Surg Neurol Int 2012;3:23, 7806.92939. Epub 2012 Feb 15.

[11]   Colombo N, Berry I, Kucharczyk J, Kucharczyk W, de Groot J, Larson T, et al. Posterior pituitary gland: appearance on MR images in normal and pathologic states. Radiology 1987;165(2):481–5.

[12]   Fujisawa I, Kikuchi K, Nishimura K, Togashi K, Itoh K, Noma S, et al. Transection of the pituitary stalk: development of an ectopic posterior lobe assessed with MR imaging. Radiology 1987;165(2):487–9.

[13]   Ioachimescu AG, Hamrahian AH, Stevens M, Zimmerman RS. The pituitary stalk transection syndrome: Multifaceted presentation in adulthood. Pituitary 2012;15(3):405–11.

[14]   Prayer D, Grois N, Prosch H, Gadner H, Barkovich AJ. MR imaging presentation of intracranial disease associated with Langerhans cell histiocytosis. AJNR Am J Neuroradiol 2004;25(5):880–91.

[15]   Anastasilakis AD, Kaltsas GA, Delimpasis G, Wilkens L, Kanakis G, Makras P. Distinctive growth pattern in a patient with a delayed diagnosis of Langerhans' cell histiocytosis. Pituitary 2012;15(Suppl. 1):S28–32.

[16]   Yong TY, Li JY, Amato L, Mahadevan K, Phillips PJ, Coates PS, et al. Pituitary involvement in Wegener's granulomatosis. Pituitary 2008;11(1):77–84.

[17]   Li Y, Zhang Y, Xu J, Chen N. Primary pituitary lymphoma in an immunocompetent patient: a rare clinical entity. J Neurol 2012;259(2):297–305.

[18]   Vates GE, Berger MS, Wilson CB. Diagnosis and management of pituitary abscess: a review of twenty-four cases. J Neurosurg 2001;95(2):233–41.

[19] Walia R, Bhansali A, Dutta P, Shanmugasundar G, Mukherjee KK, Upreti V, et al. An uncommon cause of recurrent pyogenic meningitis: pituitary abscess. BMJ Case Rep 2010;2010. Available from: http://dx.doi.org/10.1136/bcr.06.2009.1945.

[20] Morani A, Parmar H, Ibrahim M. Teaching neuroimages: sequential MRI of the pituitary in Sheehan syndrome. Neurology 2012;78(1):e3.

Index

Note: Page numbers followed by "*f*" refer to figures.

A

Abscess, pituitary, 40, 41*f*
Adenoma. *See* Pituitary adenoma
Adrenocorticotropic hormone (ACTH), 2, 15
  -dependent cyclic Cushing syndrome, 15–16
Anatomic variations, pituitary/sellar, 30–32,
  30*f*, 31*f*
Anatomy of normal pituitary gland, 2, 2*f*
Aneurysm, 33–34, 33*f*
Anterior pituitary gland, 2, 7, 26, 32, 45
  T1-weighted sagittal image without
    contrast, 6*f*
Antidiuretic hormone, 2
Apoplexy, 42–45, 44*f*
  hemorrhage, 42–45
  infarction (non-hemorrhagic), 45
Arachnoid cyst, 18, 18*f*
Atypical adenoma, 15, 15*f*

B

Blood appearance
  on T1-weighted imaging, 43
  on T2-weighted imaging, 43

C

Cavernous sinuses, 2, 10–11
  invasion of, 13, 13*f*
    pituitary adenoma with, 13, 13*f*, 14*f*
  T1-weighted coronal image, 5*f*
Cerebrospinal fluid (CSF), 2–4
  coronal T1-weighted image without
    contrast, 3*f*
  coronal T2-weighted image, 3*f*
Chordoma, 22–23, 22*f*, 23*f*
Clivus
  T1-weighted sagittal image without
    contrast, 6, 6*f*, 23
Computed tomography (CT) of sella, 1, 22,
  47–48, 47*f*

Coronal T1-weighted image without contrast,
  3*f*, 5*f*
Coronal T2-weighted image, 3*f*
Craniopharyngioma, 26–27, 26*f*, 27*f*, 48*f*
CT sella, 47–48, 47*f*
Cystic lesions, 17–20
  arachnoid cyst, 18, 18*f*
  dermoid cyst, 19–20, 19*f*
  epidermoid cyst, 19–20, 20*f*
  Rathke cleft cyst, 17–18, 17*f*
Cystic pituitary macroadenoma, 14, 14*f*, 44*f*,
  45

D

Dermoid cyst, 19–20, 19*f*
Dexamethasone, 32
"Dural tail" 27–28

E

Ectopic adenoma, 15–16, 16*f*
Empty sella, 23–25, 23*f*, 24*f*, 41–42
  partial, 25, 25*f*
  with sellar remodelling, 42*f*
Epidermoid cysts, 19–20, 20*f*

F

Fat pad, visible, 41, 41*f*
"Flow void" 33–34, 33*f*
Follicle-stimulating hormone
  (FSH), 2

G

Germinoma, 25, 26*f*
Giant invasive prolactinoma, patient with,
  12–13, 12*f*
Gonadotropins, 2, 6
Granulomatosis, of Wegener, 37, 37*f*
Growth hormone (GH), 2, 32

H

Height of pituitary gland, 4–6
Hemochromatosis, 38, 38*f*, 39*f*
Hemorrhage, 42–45
Hydrocortisone, 32
Hyperplasia, pituitary, 29–30, 29*f*
Hypophysitis, lymphocytic, 34–35, 34*f*

I

Infarction (non-hemorrhagic), 45, 45*f*
Infection, pituitary, 40, 41*f*
Infiltrative disorders, 34–38
  hemochromatosis, 38, 38*f*, 39*f*
  Langerhans cell histiocytosis, 35, 36*f*
  lymphocytic hypophysitis, 34–35, 34*f*
  neurosarcoidosis, 36, 36*f*
  Wegener's granulomatosis, 37, 37*f*
Internal carotid arteries, 2, 13, 25
  T1-weighted coronal image without
    contrast, 5*f*
Interpreting MR imaging, 1–6, 3*f*, 4*f*

L

Langerhans cell histiocytosis, 35, 36*f*
Large invasive pituitary macroadenoma, 11,
  12*f*
Lipocytes, 19
Luteinizing hormone (LH), 2
Lymphocytic hypophysitis, 34–35, 34*f*

M

Macroadenoma, 7–9, 7*f*, 9*f*
  cystic pituitary, 14, 14*f*, 44*f*
  large invasive, 11, 12*f*
  large invasive, 11, 12*f*
  T1-weighted postcontrast coronal image, 9*f*
  with mild superior displacement on optic
    chiasm, 10–11, 11*f*
  with mild superior displacement on optic
    chiasm, 10–11, 11*f*
  with stalk deviation, 10*f*
  with stalk deviation, 9–10, 10*f*
  with stalk deviation, 9–10, 10*f*
Magnetic resonance imaging (MRI), 1
Meningioma, 27–28, 28*f*
Metastases, 39, 39*f*
Microadenoma, 7, 7*f*, 9*f*

N

Nasal conchae, appearance of, 3
Neurosarcoidosis, 36, 36*f*

Non-hemorrhagic pituitary infarction, 45, 45*f*
Normal pituitary gland, 2, 2*f*, 4–6, 5*f*

O

Optic chiasm, 2, 8, 17, 25
  pituitary macroadenoma with mild superior
    displacement on, 9*f*, 10–11, 11*f*
  T1-weighted coronal image without
    contrast, 5*f*
  T1-weighted sagittal image without
    contrast, 6*f*
Outer table of the calvarium
  T1-weighted sagittal image without
    contrast, 6*f*
Oxytocin, 2

P

Partial empty sella, 25, 25*f*
Pituicytoma, 21, 21*f*, 22*f*
Pituitary adenoma, 7
  atypical, 15, 15*f*
  ectopic, 15–16, 16*f*
  macroadenoma. *See* Macroadenoma
  microadenoma. *See* Microadenoma
  with cavernous sinus invasion, 13, 13*f*, 14*f*
Pituitary hyperplasia, 29–30, 29*f*
Pituitary stalk, 4, 25
  T1-weighted coronal image without
    contrast, 5*f*
  T1-weighted sagittal image without
    contrast, 6*f*
  transection, 32–33, 32*f*
Posterior pituitary gland, 2
  T1-weighted sagittal image without
    contrast, 6*f*
Postoperative pituitary imaging, 41–42
  sellar remodeling post-pituitary surgery,
    41–42
  visible fat pad, 41, 41*f*
Post-pituitary surgery, sellar remodeling,
  41–42
Pre- and postcontrast images, identifying, 3,
  4*f*
Pregnancy, pituitary height during, 6
Primary pituitary lymphoma, 40, 40*f*
Prolactin (PRL), 2
  elevations, 13
Prolactinoma, giant invasive, 12–13, 12*f*

R

Rathke cleft cyst, 17–18, 17*f*

S

Sebaceous glands, 19
Sella, CT, 47–48, 47*f*
Sella anatomy on MRI, 4–6, 5*f*
Sella turcica, 2, 5*f*, 31–32
Sellar anatomic variations, 30–32
Sellar mass, 8, 17, 26–27, 39–40, 47*f*
Sellar remodeling post-pituitary surgery,
    41–42
Size of pituitary gland, 4–6
Snowman configuration, 25
Sphenoid sinus, 11
    T1-weighted coronal image without
        contrast, 5*f*
    T1-weighted sagittal image without
        contrast, 6, 6*f*
Stalk, pituitary. *See* Pituitary stalk
Stalk deviation, pituitary macroadenoma with,
    9–10, 10*f*
Suprasellar cistern, 2, 5*f*, 22

T

T1 hyperintensity, 19–20
T1/T2-weighted images
    appearance of CSF, 2–4
    differentiation of fat and water, 2–4
    identification of, 2–4
T1-weighted coronal image postcontrast, 3
    appearance of nasal conchae, 4*f*
    arachnoid cyst, 18*f*
    atypical adenoma, 15*f*
    chordoma, 22*f*
    cystic pituitary macroadenoma, 14*f*
    dermoid cyst, 19*f*
    empty sella, 23*f*
    germinoma, 26*f*
    giant invasive prolactinoma, 12*f*
    hemochromatosis, 39*f*
    infarction (non-hemorrhagic), 45*f*
    Langerhans cell histiocytosis, 36*f*
    large invasive pituitary macroadenoma, 12*f*
    lymphocytic hypophysitis, 34*f*
    medial deviation of carotid arteries, 30*f*
    meningioma, 28*f*
    moderately thickened stalk, 36*f*
    neurosarcoidosis, 36*f*
    non-hemorrhagic pituitary infarction, 45*f*
    pituicytoma, 21*f*
    pituitary abscess, 41*f*
    pituitary adenoma with cavernous sinus
        invasion, 13*f*
    pituitary hyperplasia, 29*f*

pituitary macroadenoma, 9*f*
    with mild superior displacement of the
        optic chiasm, 11*f*
    with stalk deviation, 10*f*
pituitary metastasis, 39*f*
pituitary microadenoma, 7*f*
primary pituitary lymphoma, 40*f*
Rathke cleft cyst, 17*f*
stalk thickening upon reassessment, 35*f*
vascular lesions (aneurysm), 33*f*
Wegener's granulomatosis, 37*f*
T1-weighted coronal image precontrast
    atypical adenoma, 15*f*
    chordoma, 22*f*
    craniopharyngioma, 26*f*
    empty sella, 23*f*, 24*f*
    empty sella with sellar remodeling, 42*f*
    germinoma, 26*f*
    hemorrhage, 43*f*
    large invasive pituitary macroadenoma, 12*f*
    medial deviation of carotid arteries, 30*f*
    meningioma, 28*f*
    nasal conchae, 4*f*
    neurosarcoidosis, 36*f*
    partial empty sella, 25*f*
    pituicytoma, 21*f*
    pituitary hyperplasia, 29*f*
    pituitary macroadenoma, 9*f*
        with mild superior displacement of the
            optic chiasm, 11*f*
        with stalk deviation, 10*f*
    pituitary metastasis, 39*f*
    pituitary microadenoma, 7*f*
    pituitary stalk transection, 32*f*, 33
    Rathke cleft cyst, 17*f*
    vascular lesions (aneurysm), 33*f*
    Wegener's granulomatosis, 37*f*
T1-weighted coronal image without contrast,
    5*f*
    anatomic variations, 30*f*
    cystic pituitary macroadenoma, 44*f*
    ectopic pituitary adenoma, 16*f*
    pituitary apoplexy, 44*f*
    visible fat pad, 41*f*
T1-weighted sagittal image postcontrast
    appearance of nasal conchae, 3, 4*f*
    arachnoid cyst, 18*f*
    ectopic adenoma, 15–16, 16*f*
    ectopic bright spot/ectopic posterior
        pituitary, 31*f*
    empty sella, 24*f*
    epidermoid cyst, 20*f*
    hemochromatosis, 38*f*

T1-weighted sagittal image postcontrast
(*Continued*)
  infarction (non-hemorrhagic), 45*f*
  Langerhans cell histiocytosis, 36*f*
  large invasive pituitary macroadenoma, 12*f*
  lymphocytic hypophysitis, 35*f*
  meningioma, 28*f*
  neurosarcoidosis, 36*f*
  patient with giant invasive prolactinoma,
    12*f*
  pituicytoma, 21*f*
  pituitary macroadenoma, 9*f*
    with mild superior displacement of the
      optic chiasm, 11*f*
    with stalk deviation, 10*f*
  pituitary metastasis, 39*f*
  pituitary microadenoma, 7*f*
  pituitary stalk transection, 32*f*
  primary pituitary lymphoma, 40*f*
  Wegener's granulomatosis, 37*f*
T1-weighted sagittal image precontrast
  chordoma, 23*f*
  craniopharyngioma, 26*f*
  dermoid cyst, 19*f*
  ectopic bright spot/ectopic posterior
    pituitary, 31*f*
  empty sella, 24*f*
  empty sella with sellar remodeling, 42*f*
  Epidermoid cyst, 20*f*
  hemochromatosis, 38*f*
  hemorrhage, 43*f*
  large invasive pituitary macroadenoma, 12*f*
  lymphocytic hypophysitis, 34*f*
  meningioma, 28*f*
  nasal conchae, 4*f*
  neurosarcoidosis, 36*f*
  partial empty sella, 25*f*
  Pituicytoma, 21*f*
  pituitary hyperplasia, 29*f*
  pituitary macroadenoma, 9*f*

    with mild superior displacement of the
      optic chiasm, 11*f*
    with stalk deviation, 10*f*
  pituitary microadenoma, 7*f*
  pituitary stalk transection, 32*f*
  Wegener's granulomatosis, 37*f*
T1-weighted sagittal image without contrast,
  6*f*
  apoplexy, 44*f*
  visible fat pad, 41*f*
T2-weighted axial image
  arachnoid cyst, 18*f*
  epidermoid cyst, 20*f*
T2-weighted coronal image
  craniopharyngioma, 27*f*
  cystic pituitary macroadenoma, 14*f*
  cystic pituitary macroadenoma, 44*f*
  empty sella with sellar remodeling, 42*f*
  hemochromatosis, 39*f*
  microadenoma on, 7*f*
  pituicytoma, 22*f*
  pituitary apoplexy, 44*f*
  pituitary macroadenoma, 9*f*
  pituitary microadenoma, 7*f*
  Rathke cleft cyst, 17*f*
  vascular lesions (aneurysm), 33*f*
  Wegener's granulomatosis, 38*f*
Teratoma, 19
Thyroid-stimulating hormone, 2
Time-of-flight MRA, 5*f*, 33*f*
Transection, of pituitary stalk, 32−33, 32*f*

V

Vascular lesions (aneurysm), 33−34, 33*f*
Visible fat pad, 41, 41*f*
Volume averaging, 46, 46*f*

W

Wegener's granulomatosis, 37, 37*f*

Printed in the United States
By Bookmasters